I0454450

HOW TO GROW PEONIES

The beginners guide to growing, caring and harvesting peonies at home and garden plus beautiful varieties

Larry Pat

Copyright © by Larry Pat 2023. All rights reserved.

Before this document is duplicated or reproduced in any manner, the publisher's consent must be gained. Therefore, the contents within can neither be stored electronically, transferred, nor kept in a database. Neither in Part nor full can the document be copied, scanned, faxed, or retained without approval from the publisher or creator.

TABLE OF CONTENT

INTRODUCTION

Peonies, beloved for their stunning, voluminous blooms, are perennial flowers that thrive when planted in the autumn, gracing gardens with their beauty throughout spring and summer. Available in a range of colors, from soft pinks to vibrant reds, they are regarded as the "queen of flowers" and evoke a sense of romance. Growing and caring for these flowers involves attention to detail but promises increasingly exquisite blossoms with each passing year. Known botanically as Paeonia spp., peonies are sun-loving plants with a preference for neutral soil pH. Blooming in spring and summer, they come in various colors, including pink, purple, red, white, and yellow, and are hardy in zones 3 to 8.

About Peonies

Peonies, perennial wonders, make a recurring appearance each year, captivating with their large, luxurious flowers and vibrant green foliage. In the 1930s and '40s, plant catalogs featured only three peony options—white, crimson, and rose pink. However, the present day boasts an extensive array of thousands of peony varieties.

Not only do peonies showcase delightful foliage, but they are also uncomplicated to cultivate, serving as magnificent cut flowers. Resilient in diverse climates, they thrive in colder regions where a chilly period is essential for bud formation, posing a challenge for some gardeners in warmer southern areas.

Once these plants find their place in the garden, they establish themselves for generations, with some peonies flourishing for an impressive century.

Types of Peonies

Originating in ancient China around 1000 B.C., peonies boast a rich history with over 6,500 varieties. Among this vast assortment, each peony can be categorized into one of three main types:

1. Herbaceous: This prevalent and cold-hardy peony type is characterized by its growth on stems.

2. Tree: Represented by stemmed shrubs, tree peonies offer a distinct variation within the peony family.

3. Itoh: A hybrid category, Itoh peonies result from a cross between herbaceous and tree peonies, combining characteristics of both parent types.

When Is Peony Season? When Do Peonies Bloom?

The blooming period of peonies spans from late spring to early summer, with the specific timing influenced by both the geographical location and the particular peony variety being cultivated.

Nurseries commonly offer a range of peony varieties classified as early, midseason, or late bloomers. This diversity allows enthusiasts to extend the enjoyment of peony blossoms over several weeks, maximizing the duration of the peony season.

Peonies are robust and can withstand harsh conditions, thriving in climates as cold as Zone 3 and extending their range down to the warmer Zones 7 and 8. The key to successful cultivation is relatively straightforward across most of the U.S.: provide ample full sun exposure and well-drained soil. Notably, peonies appreciate colder winters as they require a chilling period for optimal bud formation.

Types of Peony Flowers

Peonies offer a diverse array of flower types, providing enthusiasts with six distinctive options to choose from. These include:

1. Anemone: Featuring a central cluster of narrow petals surrounded by a ring of larger, petal-like structures.

2. Single: Consisting of a single row of petals encircling the stamens and pistils, showcasing a simple yet elegant appearance.

3. Japanese: Characterized by a central cluster of petaloids, or narrow petals, often surrounding a ring of contrasting petal-like structures.

4. Semi-Double: Combining characteristics of both single and double varieties, semi-double peonies feature a mix of single and multiple rows of petals.

5. Double: Abundant and opulent, double peonies boast multiple layers of petals, creating a lush and full appearance.

6. Bomb: Resembling a tight ball, bomb peonies have a dense cluster of petals that completely envelop the stamens and pistils.

In addition to their varied flower types, peonies also offer a range of fragrances. Some, like 'Festiva Maxima' and 'Duchesse de Nemours,' emit captivating rose-like scents, while others feature lemony fragrances or may have no scent at all. This diversity allows for a personalized selection based on both visual and olfactory preferences.

Where to Plant Peonies

Peonies serve as elegant sentinels, enhancing walkways or forming charming low hedges. Following their striking bloom, these plants maintain a robust presence with a dense clump of glossy green leaves that endure throughout the summer. Come fall, the foliage undergoes a transformation, turning purplish-red or gold, presenting a stately and dignified appearance akin to flowering shrubs.

In mixed borders, peonies harmonize beautifully with companion plants such as columbines, baptisias, and veronicas. They also make excellent pairings with irises and roses, creating visually appealing combinations. For instance, planting white peonies alongside yellow irises and a delicate haze of forget-me-nots produces a delightful ensemble.

Similarly, pink peonies can be complemented with the vivid hues of blue Nepeta or violets, showcasing the versatility of peonies in enhancing a variety of garden settings.

DIFFERENT WAYS TO GROW PEONY PLANTS

Growing From Seed

While it's possible to grow peonies from seed, it's not the most efficient method for introducing this beautiful plant to your yard. The process can be time-consuming, taking anywhere from five to seven years for a peony to start producing flowers from seed. To enjoy the blossoms more quickly, it is advisable to opt for purchasing a potted plant from a root propagation.

It's important to note that unless the parent plant is an open-pollinated variety, the seeds may not grow true. This means that the resulting plants could differ in appearance from the parent plant.

However, if you already have peonies in your garden and wish to try planting from seeds produced by your own plants, you can follow a few steps. Begin by collecting the seed pods in late summer or early fall once they have turned brown and split open. After collecting the seeds, place them in lukewarm water and discard any that float, as they are less likely to sprout. Plant the remaining seeds promptly in an area receiving at least six hours of sunlight daily, ensuring well-draining soil that is kept consistently moist but not overly wet. This method allows you to experiment with growing peonies from your existing plants, but it requires patience and may yield variations in the resulting plants.

Mature Potted Plants

The most straightforward and effective way to grow peonies is to use bare-root stock, which is dug and shipped during their dormant period in the fall, ensuring optimal blossom production.

When planting multiple peonies, make sure you have ample space, allowing at least four feet between each plant. Dig a hole that is 2 to 3 inches deep in a sunny location, and choose a spot sheltered from strong winds, as the weight of the flowers may cause the plant to bend during storms. In regions with harsh weather, consider using a support tower for added stability. Once planted, it typically takes about three years for peonies to showcase a profusion of flowers.

Transplanting

Peonies, once established, do not take well to transplanting, as the process can lead to a prolonged recovery period and a potential delay in blooms for up to three years. Transplanting can be particularly risky, causing severe damage and requiring an extended recovery if not executed properly.

The reason lies in the deep and extensive root systems of peonies, which form strong connections with the soil over time.

Young peony plants may initially produce smaller flowers, but as they mature and develop robust root systems, the blooms tend to increase in both size and quality. Transplanting disrupts this natural progression.

If you find it necessary to transplant peonies, adhere to these guidelines:

1. Choose early fall or early spring for transplanting.

2. Dig carefully, taking precautions to avoid damage to the root system.

3. Select a well-draining location with either sunlight or partial shade.

4. Place the peony at the appropriate depth (2 to 3 inches) with buds facing upward, then backfill with soil.

5. Water thoroughly and apply organic mulch to retain moisture and prevent weed growth.

PEONIES REQUIREMENTS

Temperature

Peonies thrive in ample sunlight and require a robust winter freeze to store energy in their roots for the following spring. Consequently, they may not fare well in warmer zones such as Southern California, Florida, and certain southern states.

Light

For optimal growth and the production of large, healthy flowers, peonies need a minimum of six hours of full sunlight.

Soil

Plant peonies in well-draining soil with a slightly acidic to neutral pH, enriched with organic compost. Avoid planting the root too deep to ensure proper blooming. Cover the buds with only one to two inches of soil after positioning the root in the hole.

Fertilizer

After blooming, apply a mix of compost and a small amount of fertilizer to the base of the plant annually.

Water

Supply 1 to 2 inches of water per week. While peonies don't require excessive irrigation, consistent watering is essential.

Pruning

Prune peonies by cutting them back to ground level in late fall or early spring before new growth emerges. During pruning, check for and remove any dead or damaged stems to maintain plant health.

Deadheading

When cutting flowers, leave at least two sets of leaves on the plant. Deadhead peonies as they bloom, ensuring the leaves are retained to support the plant's energy storage for the following year.

Staking

Once established, peonies can flower abundantly for many years. Some varieties, especially double-flowered types with large, heavy blooms, may require staking. Use peony cages, sturdy tomato cages, bamboo stakes, or roping to provide support and prevent stems from splaying, particularly after heavy rains. Proper support is crucial to preventing plant damage.

Planting Peonies from Containers

For those seeking the convenience of immediate results, potted peony specimens are available at local garden centers and can be planted in the spring. Enhance the planting process by incorporating a generous amount of compost and phosphate-rich fertilizer, with bonemeal being an excellent choice.

During the initial years after planting, try to resist cutting flowering stems, particularly in the first, second, and even third year. This restraint allows the plant to accumulate the necessary energy for robust establishment and prolific growth in the years to follow. Although it might be tempting to indulge in cutting blooms, the patience invested will yield a fully mature plant that rewards with numerous healthy blooms over time.

While potted peonies provide instant gratification, planting bare-root tubers in the fall offers even better results.

Not only does this method deliver superior outcomes, but it is also a more economical approach, considering that the cost of bare-root tubers is a fraction of potted peony plants.

How to Plant Bare-Root Peonies

Dig a hole that is 2-3 times wider than the tuber. Enhance the soil by adding compost to ensure proper drainage. Introduce a phosphate-rich fertilizer like bone meal and a handful of mycorrhizae fungi to encourage robust root growth.

Place the tubers just below the soil surface, with the eyes (red growth buds) around 1 inch below. It's crucial not to plant the tubers too deeply, as this can affect flowering. Space the roots approximately three feet apart to accommodate the plant's mature size. In early spring, before foliage emerges, lightly sprinkle high-phosphorus fertilizer and no more than 1 inch of compost on the soil.

This provides nutrients for the growing season, supporting bud and bloom development.

PLANTING

Peonies, while not excessively demanding, require thoughtful consideration when choosing a location due to their aversion to disturbance and poor tolerance for transplantation.

Optimal conditions for peonies involve full sun exposure, with the best blooming results achieved in a sunny spot receiving 6 to 8 hours of sunlight daily. In regions with intense sunlight, providing some shade, particularly in southern states, is advisable.

To prevent the risk of toppling from strong winds, offer shelter to peonies, as their large blooms can make them top-heavy. Staking may be necessary for additional support. Avoid planting too close to trees or shrubs, as peonies prefer not to compete for essential resources such as food, light, and moisture.

Cultivate peonies in deep, fertile, humus-rich, and well-draining soil that retains moisture. It is essential for the soil to have a neutral pH to create favorable conditions for peony growth.

When to Plant Peonies

Maintaining peony plants is relatively undemanding, given proper planting and establishment. It's crucial to note, however, that peonies are not amenable to transplanting, so careful consideration of the planting site is essential.

For optimal results, plant peonies in the fall, specifically in late September and October in most regions of the U.S. In Zones 7 and 8, where fall extends later, planting can be done even later in the season. If there's a need to relocate a mature plant, fall is the recommended time, specifically when the plant has entered dormancy.

To ensure successful establishment, peonies should be settled into their designated location approximately six weeks before the ground freezes. While planting peonies in the spring is feasible, it's worth noting that spring-planted peonies tend not to perform as well. Expert consensus suggests that they generally exhibit a delay of about a year compared to those planted in the fall.

How to plant peonies

Peonies are typically available as bare-root tubers, containing 3 to 5 eyes (buds), and are divisions from a 3- or 4-year-old plant. When planting, it's important to create sufficient space between peonies, around 3 to 4 feet apart, to promote effective air circulation and prevent the development of diseases in stagnant, humid conditions.

Prepare a well-drained, sunny spot for planting by digging a generously sized hole, approximately 2 feet deep and 2 feet across.

Enhance the soil with organic material in the planting hole, especially if it is heavy or sandy, by incorporating extra compost. Introduce about one cup of bone meal into the soil to provide essential nutrients.

Position the root in the hole, ensuring that the eyes face upward on top of a soil mound, with the roots just 2 inches below the soil surface. It's crucial not to plant too deeply. For southern states, especially with early-blooming varieties, plant them about 1 inch deep and provide some shade.

Backfill the hole, taking care to prevent soil settling that might bury the root deeper than 2 inches, and gently tamp the soil. When planting a peony from a container, ensure not to bury it deeper than its original depth in the pot. Water thoroughly at the time of planting to promote a successful start for the peony.

GROWING

How to Care for Peonies

Similar to children, young peonies require time to establish themselves, bloom, and grow. Unlike most perennials, peonies don't demand frequent digging and dividing every few years. They thrive with a hands-off approach, flourishing under benign neglect.

Avoid excessive use of fertilizer. Prior to planting, prepare the soil thoroughly by incorporating compost and a modest amount of fertilizer. This initial effort should suffice. If the soil is lacking, apply fertilizer such as bone meal, compost, or well-rotted manure in early summer after the peonies have bloomed and the flowers have been deadheaded. It's advisable not to fertilize more frequently than every few years.

Provide structural support for peony stems, especially if they display any weakness unable to bear the weight of their large blossoms. Consider using three-legged metal peony rings or wire tomato cages, allowing the plant to grow through the center of the support.

Regularly deadhead peony blossoms as soon as they start to fade. Trim to a robust leaf to ensure that the stem doesn't protrude from the foliage. This practice contributes to the overall health and appearance of the peony plant.

Fall Peony Care

Once peony leaves start to fade, provide a side-dressing of a balanced, slow-release fertilizer, but avoid using fertilizers high in nitrogen.

After the first frost, when the foliage has completely withered, cut the peony plant to the ground in the fall to prevent overwintering diseases.

Avoid excessive mulching; in regions with severe cold temperatures, for the initial winter after planting, use pine needles or shredded bark as mulch, but keep it very loose. Remove the mulch in the spring.

While peonies generally don't require division, fall presents a suitable time for division or transplantation if the plants have become too large. Remove the leaves, then dig around the plant's roots in a large circle and lift it. If dividing, ensure that each new section has at least 3 to 5 eyes before replanting.

Recommended Varieties

Peonies grace gardens with their blooms from late spring to early summer, and you can strategically plan for a continuous floral display from mid-May to late June by incorporating various varieties. Here are some choices:

1. **Anemone (advanced Japanese form):** 'Laura Dessert' - Cream/pale lemon to white; strong fragrance; early-season.

2. **Single (one row of petals):** 'Sparkling Star' - Deep pink; early- to midseason.

3. **Japanese (decorative centers):** 'Carrara' - White, with soft yellow centers; fragrant; midseason.

4. **Semi Double (five or more guard petals and prominent centers):** 'Pink Hawaiian Coral' - Pink to coral rose form; slight fragrance; early-season.

5. Double (large petals, needs support): 'Candy Stripe' - White, with red streaks; slight fragrance; mid- to late-season.

6. Bomb (large, like a scoop of petals): 'Angel Cheeks' - Soft pink; slight fragrance; midseason.

7. Itoh Peony: Named after its developer, Japanese botanist Dr. Toichi Itoh, this flower is a cross between a herbaceous (bush) peony and a tree peony. Its large single, semi-double, and double blooms appear later in the season and last longer than traditional peonies.

HARVESTING

Peonies are excellent as cut flowers and can last for over a week in a vase. For optimal results, cut long stems in the morning when the buds are still relatively tight. After cutting, wrap the peony stems in a damp paper towel and store them in a plastic bag in the refrigerator until you're ready to use them. When taking them out, give the stems a fresh cut and place them in lukewarm water to rejuvenate the flowers.

Wit and wisdom

The ornamental peony, cultivated and bred in China for over 2,000 years, made its introduction to Europe and America around 1800. North America hosts two native peony species, namely Brown's (also known as western peony) spanning from California to Montana, and the California peony found along the Pacific coast.

Peonies are symbolic of a happy life and marriage. Marco Polo vividly described peony blossoms as "roses as big as cabbages." In the ancient practice of phenology, the blossoming of peonies signals a safe time to plant heat-loving melons like cantaloupe.

Notably, peony petals are edible and can be incorporated into summer salads or used as a decorative garnish for beverages like lemonade and iced tea.

Pest/Diseases

Peonies are typically robust and resilient, being hardy plants. Additionally, they fall into the category of deer-resistant plants for your garden.

Nevertheless, they can be vulnerable to various issues such as:
- *Verticillium wilt*
- *Ringspot virus*
- *Tip blight*

- *Stem rot*
- *Botrytis blight*
- *Leaf blotch*
- *Japanese beetles*
- *Nematodes*
- *Aphids*

Why Are There Ants on My Peonies?

A common curiosity among gardeners is the presence of ants on peony buds. There's no need for concern, as these ants are simply consuming the peony's nectar. In return, they actively combat bud-eating pests. The ants are drawn to the sugary droplets on the exterior of the flower buds or to the honeydew produced by scale insects and aphids. It's advised not to spray the ants, as they play a helpful role in safeguarding your peonies.

COMMON PROBLEMS WITH PEONIES

Once established, peonies are a hardy shrub that demands minimal maintenance. However, like any plant, they are susceptible to specific issues that become apparent when they need extra attention. Look out for the following signs:

1. Leggy Stems

If your peonies are producing leggy stems and few flowers, it indicates insufficient sunlight. To address this, consider relocating the plant to a sunnier spot, which might take a few seasons to restore blooming. Trimming trees or shrubs that cast shade on the peonies is an alternative solution.

2. Crispy Leaves

Singed or burnt leaves suggest that your peonies are receiving too much sun, especially during long and hot summer days. Instead of moving them to a shadier spot, provide extra shade using structures or other plants around the peonies.

3. No Blooms

If your peonies have abundant greenery but lack blooms, it could be a sign of planting them too deep in the soil. Excessive soil around the plant's crown may hinder blooming. To address this, dig up and replant the peonies, and they should start blooming in subsequent seasons.

4. Powdery Mildew

Peonies may encounter powdery mildew, affecting the appearance of their leaves. Ensuring proper air circulation, avoiding overhead watering, and treating them with neem oil or a homemade baking soda spray can help resolve this issue.

SEASON-BY-SEASON PEONY CARE CHART

Consult our convenient season-by-season peony care chart to ensure you provide the necessary attention to these bushes throughout the year:

Spring

- In late winter or early spring, apply a thin layer of compost around peony bushes.
- Once a new foot of growth is established by mid-spring, use a slow-release fertilizer to enhance nutrient uptake.
- Install support structures around top-heavy established plants, such as trimmed-down old tomato cages, to keep blooms upright.

Summer

- Prune peonies throughout the summer growing season, focusing on maintaining shape and controlling insects.

- Be cautious while pruning to avoid cutting back stems containing the large round bulbs that produce the signature fluffy blooms.

Fall

- After the first frost in October or November, perform a deep prune by cutting peony bushes down to soil level to stimulate new, healthy growth in spring.
- This season is suitable for moving plants or taking cuttings for propagation. Dig up roots, leaving them untouched for a couple of days to soften before cutting.
- Transplant peonies sparingly, as they don't prefer frequent moves.

Winter

- Peonies do not require care during winter as their bulbs remain underground in cold weather.

BONUS: 20 BEAUTIFUL PEONIES VARIETIES TO GROW

Peonies come in various forms, including herbaceous, tree, and intersectional peonies, each with distinct characteristics. Here's a list of specific peony varieties along with their features:

1. Peony 'Coral Charm'

- Flowers in June, changing from salmon pink to orange and finally yellow.
- Height and Spread: 85cm x 85cm.

2. Peony 'Duchess of Kent'

- Vigorous Japanese tree peony with ruby tulip-shaped buds.
- Height and Breath: 1.2m x 90cm.

3. Peony 'Duchess of Marlborough'

- Famous Japanese tree peony with large, pale pink blooms.

- Height and Spread: 1.2m x 90cm.

4. Peony 'Hillary'

- Unique apricot and magenta two-tone blooms with a spicy fragrance.
- Intersectional peony with a 'bush' habit.
- Height and width: 90cm x 90cm.

5. Peony 'Laura Dessert'

- Fragrant French cultivar with pale creamy-yellow blooms.
- Height and width: 90cm x 75cm.

6. Peony 'Lollipop'

- Intersectional peony with pink-speckled apricot blooms.
- Mid-season flowering in a 'bush' form.
- Height and breadth: 70cm x 90cm.

7. Peony 'Pink Hawaiian Coral'

- Early-season flowering with large coral-pink blooms.

- Vigorous and easy to grow, reaching around 90cm in height.
- Height and width: 90cm x 75cm.

8. Peony 'Prairie Moon'

- Herbaceous peony with white single blooms and a bright yellow center.
- Early-flowering variety, adding interest to the border in June.
- Height and width: 85cm x 75cm.

9. Peony 'Shima Nishiki'

- Eye-catching red and white striped Japanese tree peony.
- Forms a neat 'shrub' habit.
- Height and width: 1.2m x 90cm.

10. Peony 'Souvenir de Maxime Cornu'

- French tree peony with very large, double ruffled peach-colored blooms.
- Blooms hang downwards with mature color strengthening.
- Height and width: 1.2m x 90cm.

11. Peony 'Bartzella'

- Intersectional peony with large, frilled, yellow flowers and a citrus fragrance.
- Occasional second flush of flowers in September.
- Height and width: 90cm x 90cm.

12. Peony 'Chocolate Soldier'

- Cup-shaped flowers in the darkest red with a central flush of bright yellow stamens.
- Works well with other dark-flowered plants and makes an excellent cut flower.
- Height and width: 70cm x 70cm.

13. Peony 'Nippon Beauty'

- Late-flowering herbaceous peony with deep red, semi-double flowers.
- Cream-flushed central ball of fringed petals.
- Height and width: 90cm x 75cm.

14. Peony 'Claire de Lune'

- Pale lemon-yellow, single flowers with a huge center of deep yellow stamens.
- Gently fragrant with red flower stems and rich green leaves.
- Height and breadth: 90cm x 80cm.

15. Peony 'Bowl Of Beauty'

- Incredible pink flowers with a central core of tiny, strap-like petals and a wonderful fragrance.
- Suitable for large herbaceous borders and can be grown in a large pot.
- Height and width: 80cm x 80cm.

16. Peony 'Sarah Bernhardt'

- Enormous, fragrant, double flowers in rose pink with ruffled white edges.
- Flower stems may need staking, and it makes an excellent cut flower.
- Height and breadth: 1m x 90cm.

17. Peony 'Le Cygne'

- Double-flowered peony with large, white blooms, curved petals, and yellow stamens.
- Perfect for mixed herbaceous borders and containers, making an excellent cut flower.
- Height and width: 85cm x 50cm.

18. Peony 'Immaculee'

- Bowl-shaped, white flowers with a central core of tiny, strap-like white petals and scarlet stamens.
- Good fragrance, suitable for arranging in mixed herbaceous borders or containers.
- Height and width: 90cm x 50cm.

19. Peony 'Instituteur Doriat'

- Anemone-type flowers in deepest pink with petaloids having white tips.
- Perfect for mixed herbaceous borders or containers, making an excellent cut flower.
- Height and width: 1m x 1m.

20. Peony 'Bowl of Love'

- More compact version of 'Bowl of Beauty' with pink flowers and a wonderful fragrance.

- Height and width: 70cm x 50cm.

GROWING HEALTHY PLANTS AND PREVENTING DISEASE/PEST INVASIONS

Watering Tips

Ensure even and timely watering, avoiding both over and under-watering to prevent plant stress. Direct water to the soil at the plant's base, avoiding wetting leaves to prevent fungus. Morning watering allows the plant to hydrate during the day, while night watering may invite fungus due to lingering moisture in cool temperatures. Established shrubs and perennials usually don't need regular watering once rooted.

Regularly trim away dead or dying foliage and stems to prevent rot and maintain plant health. Provide appropriate fertilizer following label instructions, assess sunlight needs (8+ hours for peonies), and control weeds using mulch and manual weeding or tools.

Daily plant checks are essential to address pests and fungus promptly, utilizing organic methods and beneficial insects as preventive and corrective measures. Combatting Botrytis Blight and Powdery Mildew involves physically removing infected plant material using clean pruners. Disinfect between plants and dispose of infected material safely, avoiding composting to prevent the spread of diseases to other plants in your yard.

HOW TO HANDLE PEONIES IN THE AUTUMN

Cutting Back Peonies

Similar to general bulb plant care, prune all old stems on herbaceous peonies in late fall after the first frost turns the foliage yellow. This signals successful energy transfer to the roots for winter storage and spring growth.

Discard cut foliage to prevent gray mold, a winter-resistant fungus that affects peonies and may survive in composted old stems.

Transplanting Peonies

Fall is optimal for digging and transplanting peonies. Carefully dig around and clear under the roots, avoiding damage to fleshy tubers. Gently lift the tuber clump with a wide spade or pitchfork, minimizing root disturbance. Transplant to a new location with full sun, well-draining, rich soil, planting just beneath the soil level. Water thoroughly.

Dividing Peonies

Fall is the ideal time for dividing and multiplying plants. Large, well-established peonies can be divided to renew growth or create new plants. Cut back foliage, carefully dig up the root system, and cut the tuber clump into sections, each with three to five eyes and several roots.

Replant each section in a new location, placing the buds 1 to 2 inches below the soil surface. Water thoroughly.

Winter Protection for Peonies

In colder hardiness zones (especially 3-5), provide winter protection for peonies. In late fall, apply a 2 to 3-inch organic mulch layer (shredded bark or straw) to insulate them and ensure a cozy winter, aiding their emergence in spring.

How and When to Cut Peonies

Picking peonies requires some preparation and basic knowledge for optimal results. Here are recommendations to follow:

Knowing When Peonies are Ready to be Cut

1. Have a bucket of room temperature, clean water ready with flower food added for immediate placement of stems.
2. When using the stems right away, provide them with an immediate drink. For delayed

use, see tips below for preserving in the fridge.

3. Preferably, cut peonies when they are in the "soft marshmallow" bud phase. When the bud is softly compressed and has a texture similar to that of a squishy marshmallow, it indicates that it is ready to be harvested.

4. If the bud feels harder than a marshmallow, let it ripen a bit more while keeping a close eye, as peony buds can transition quickly from hard to soft marshmallow in a matter of hours.

5. Picking at the soft marshmallow stage ensures longer vase life and allows the flowers to fully open.

6. Avoid picking buds too soon, as underdeveloped petals may result in a wonky-looking half-bloom. If the flowers are too blown out and open, they are still beautiful but may have a shorter vase life.

Cleaning Peony Stems

After cutting peonies from the plant, it's essential to follow these steps immediately for optimal vase life:

1. Remove all foliage and minor buds right away after cutting. Even though it might seem counterintuitive to remove potential flowers, these minor or side buds can drain energy from the main bloom, significantly reducing its vase life if left on the stem.

2. The removal of minor buds is crucial as they are not developed enough to open in water and can hinder the main bloom's longevity.

3. Remove all foliage, not only because it may rot in water (a concern for all cut flowers) but also because leaves can draw energy from the main bloom, affecting its vase life.

4. Cut off foliage and unnecessary buds immediately after cutting the blooms from the plant, before placing them in the water bucket. This ensures that the main bloom receives maximum energy for a longer vase life.

EXTRA PEONY HARVESTING TIPS

For optimal peony harvesting and vase preparation, consider the following steps:

1. Stem Length

- Cut peony stems as long as possible, around 24 inches, to provide flexibility for arranging them in the final vase. Longer stems allow for various height options.
- Perform a second cut under the water as you place the peonies in the bucket. This second cut, made under the water, ensures optimal water absorption.
- Use clean and sharp snips for cutting. Consider using quality snips like those by Felco or Fiskars. Before each use, spray or dip the snip blades in rubbing alcohol for sanitation.
- Trim the stems at a 45-degree angle to enhance water absorption.

2. Harvesting Time

- Harvest peonies in the early morning or late afternoon/evening for the best results.
- Avoid cutting them in the heat of the day, as this can stress the flowers, cause them to open too quickly, and shorten their vase life.
- Ensure that at least two sets of leaves remain on stems left on the plant during harvesting. This allows the plant to continue growing and storing food over the summer.

How to Make Your Peonies Grow Well in Vase

Peonies can maintain their freshness for 2-3 weeks with proper care. Different peony varieties open at varying rates; for example, Coral Charm, Festiva Maxima, and Karl Rosenfeld open quickly, while Sarah Bernhardt opens more slowly.

Cooler temperatures consistently extend their vase life. To slow down the opening process, store cut peonies in the refrigerator in a vase of water, removing them when ready for display. If refrigeration space is limited, bundle stems together, dry them off, and place them in a plastic bag with paper towels. Alternatively, wrap the bare stems in tissue paper or newspaper before refrigerating.

Ensure the blooms, leaves, and stems are dry before storage. If a refrigerator is unavailable, store peonies in the coldest room possible. Avoid freezing temperatures, and keep them away from moisture, fruits, and vegetables, as ethylene gas can accelerate aging. When removing peonies from the fridge, give the stems a fresh cut and place them in warm water with flower preservative. They should revive and open beautifully within 24 hours, lasting about a week in freshwater.

How to extend peonies vase life and freshness

To extend the vase life of peonies and control their opening rate, consider the following tips:

Differential Opening Rates:
Different peony varieties open at varying rates. For instance, Coral Charm, Festiva Maxima, and Karl Rosenfeld tend to open quickly, while Sarah Bernhardt opens more slowly.
Cooler temperatures consistently prolong the vase life of peonies.

Refrigeration for Slower Opening:
To slow down the opening process, store cut peonies in the refrigerator in a vase of water. Remove them when ready to display or use.
If space is limited, bunch the stems together, dry them off, and place them in a plastic bag with paper towels wrapped around the stem bases.

This method helps absorb excess moisture and keeps the stems slightly moist.

Alternatively, wrap the entire bouquet (bare stems) in tissue paper or newspaper before refrigerating. Lay the bunch flat and check daily for signs of mold, changing the paper towel if needed.

Ensure the blooms, leaves, and stems are dry before refrigerating.

Storage Without a Fridge:

If a refrigerator is unavailable, store peonies in the coldest and dimmest room possible, such as a closet, bathroom, basement, or garage. Avoid freezing temperatures; around 41 degrees Fahrenheit is ideal.

Keep peonies away from moisture, fruits, and vegetables, as ethylene gas from these items can accelerate the aging process.

Reviving Limp Flowers:

When removing peonies from the fridge, they may appear limp. Give the stems a fresh cut and immediately place them in a vase of warm water with flower preservative.

Peonies should perk up and open beautifully within the next 24 hours. They can last about a week in freshwater.

Conclusion

In this comprehensive guide, we've explored the enchanting world of peonies, delving into every aspect of their care, cultivation, and preservation. From the initial steps of planting and amending the soil to the joys of witnessing their vibrant blooms, we've covered the lifecycle of peonies.

The advice on maintenance, whether it be watering tips, dealing with pests, or the importance of proper pruning, provides readers with the knowledge to ensure these resilient flowers thrive in their gardens. Additionally, we've uncovered the art of picking and preserving peonies, revealing the secrets to extending their vase life and savoring their beauty for weeks.

Whether you're a seasoned gardener or a novice flower enthusiast, this guide serves as a valuable companion, unlocking the mysteries and celebrating the enduring allure of one of nature's most cherished blooms—the peony.

www.ingramcontent.com/pod-product-compliance
Lightning Source LLC
Chambersburg PA
CBHW062254290526
45794CB00006B/2550